Once Strangers, Forever Brothers

Once Strangers, Forever Brothers

VIETNAM – 1969-1970

Joe Carter

Deeds Publishing | Athens, GA

Published by Deeds Publishing in Athens, GA
www.deedspublishing.com

Printed in The United States of America

Cover and interior design by Deeds Publishing

ISBN 978-1-961505-50-6

Books are available in quantity for promotional or premium use.
For information, email info@deedspublishing.com.

First Edition, 2025

10 9 8 7 6 5 4 3 2 1

This book I dedicate to the men I served with, especially those who helped keep me alive over there

To Bob Babcock for reminding me to tell my story

To my wife Teresa for keeping on me to get this book written and her computer skills to help me get this to the publisher

Contents

CHAPTER I

Enlist vs Draft

Born in St. Joseph, Missouri, March of 1950, and raised on a small farm near Faucett, Missouri, I was the third oldest child of Mary and Delmar Carter. As time went on, our family grew, giving me five brothers and two sisters. We were a close-knit family with many close cousins, good neighbors, and lots of good friends.

I graduated High School in 1968. My older brother John graduated in 1966 and enlisted in the Army, so I decided since there was a high probability that I would be drafted that I would just go ahead and enlist. The Army Recruiter made it sound so good; I signed up to go in September of 1968. I went home and told mom and dad the good news, but I don't think they were too proud of me. They both went through WWII and knew I would be going to Vietnam as my brother John was already there, serving with the 101st Airborne near Hue, Vietnam.

My dad never talked about WWII until later in his life. He was a second Lieutenant on a B-17 Bomber in England. As a co-pilot, they had made eight successful bombing missions. There was a lot of fog in England and during a practice bombing mission, his plane clipped its wings with another plane and went down. Five crew members were able to bail out, but Dad's parachute did not fully open, causing him to hit the ground hard, breaking his back and knocking out his teeth.

Dad spent two years in a Military Hospital in England before they finally transferred him to Fort Riley, Kansas to fully recover.

My mom had married at eighteen, her first husband, Harold, was drafted into the Army during WWII. He was in the infantry in Italy where he was killed by hostile fire at the early age of twenty.

During the time my dad was recovering from the accident in England, he and mom had been corresponding by letters since they had known each other from high school and grew up in the same small town. Once he had fully recovered, he returned home to New Hampton, Missouri where he decided that he wanted to marry mom, and so on July 12, 1947, they were married.

CHAPTER 2

Basic Training

I left St. Joseph, Missouri by Greyhound bus on September 28, 1968, enroute to Kansas City, Missouri to take my physical, given by the Army Physicians. There were about fifty young men in the examination room. This was the beginning of losing your dignity, or shyness, because there would be no privacy from this point on like you were raised with.

After passing the physical, we were bused to Fort Leonard Wood, Missouri, arriving around midnight. A yelling Drill Sergeant greeted us. We were issued our military fatigues and housed in a two-story wooden barracks that was used in WWII. Little did we know, but our last full night of sleep was the night before we left home, and this was going to be the story of my life for the next two years.

Five o'clock the next morning came fast, we were rushed through the chow line where you gulped down your food and then sent running back to your temporary barracks where you were then loaded on trucks and taken to your assigned company for the next eight weeks. Exercising, calisthenics, marching, and strict discipline were drilled into us every day. The Army broke you down and then raised you like they wanted you to be.

In Basic training we learned how to live in all kinds of weather elements as we would bivouac in rain, snow, and the freezing weather. We spent a lot of time on the rifle range and then at night we took apart our rifles, cleaned, oiled, and reassembled them. I was trained on an M-14 semi-automatic rifle, and it was drilled into our heads that this was our best friend, take care of it, as your life will depend on it. Hand to hand combat was taught to us every day.

When the 8-week basic training was over, I was bused to the Saint Louis airport, and flown to Mc-Cord Airfield, Tacoma, Washington for advanced infantry training. (AIT) Advanced infantry training was a lot like basic training, firing rifles, lots of physical exercise, and vaccinations as well as a lot of respect and discipline.

AIT fell during the Christmas Holidays and I was a long way from home, but I was given a two-week leave. Living on Military pay of a hundred and three dollars a month did not leave me many options for going home, so I rode a Greyhound bus for three days back to Kansas City. This would make my leave time a little short, but at least I got to go home.

My brother John helped me buy a plane ticket back to Fort Lewis, Washington in January to finish my AIT training. It would be another four weeks of vaccinations and lots of physical exercise.

AIT was over in February 1969 and I was given twenty days leave to go home. This time I flew into Kansas City. Once the plane landed, I was met by my parents and, like always, it was good to be home for those twenty days. I knew on my return trip back to Fort Lewis, with orders in hand, I would soon be going to Vietnam.

Returning to Fort Lewis, Washington, all the trainees were given the rank of PFC E-3. Our dress uniforms were put into storage at Fort Lewis, and we were dressed in green Army fatigues, bused to the airport, and put on a commercial airplane. We were on our way to southeast Asia. The flight was to be fourteen hours and 7,260 miles. We stopped in Anchorage, Alaska for refueling, and then it was nonstop to Cam Ranh Bay, Vietnam.

Arriving at Cam Ranh Bay there was white sand and the temperature was 110 degrees. There was a smell of burning shit in the air. The burning of the portable toilet debris was the daily duty that every troop experienced at least once while they were in Vietnam. That was a smell you would not forget about Cam Ranh Bay or any base camp or Vietnam in general.

While I was at Cam Ranh Bay for five days, I brushed my teeth with Fluoride, learned how to dress wounds, and how to take care of yourself, and our buddies out in the field. We were taught about airing out our feet daily and drying our socks. I was given Malaria pills daily, as we later learned that in the jungle mosquitoes were terrible.

While at Cam Ranh Bay we visited the USO and were given writing paper, pens, toothpaste, along with some shaving gear for free. This was the last time we would get to spend time in civilization as we knew it for quite some time.

Vietnam was an undeveloped country, no running water, no toilets just where it was handy to go, people living off the land, no jobs, and just dirty little villages with no electricity. When the dark of night fell, it was pitch black and made it even scarier. No one ever moved around; smoking or flashlights were not advised as this would give away your position and give the enemy a target.

Sign I saw when I arrived at Cam Ranh Bay

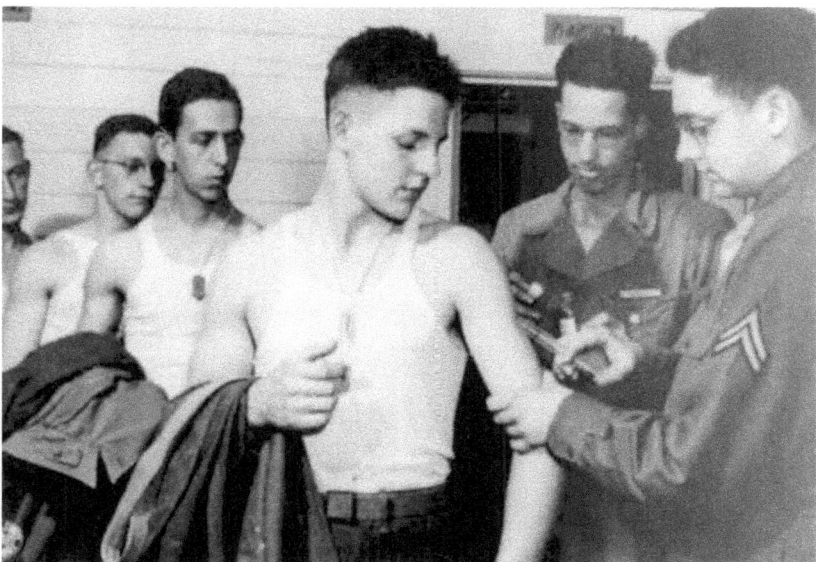

One of the many vaccinations we received before going to Vietnam.

Type of Barracks we were housed in for basic and AIT training

Bathroom privacy basic and AIT

Inspection during basic training

Mess Hall

WELCOME TO THE SUNNY TROPICS OF SOUTH VIETNAM

VIETNAM TROPICS

LIVE ENTERTAINMENT NIGHTLY

HIGHLAND RETREATS

MEET STRANGE AND INTERESTING PEOPLE

FINE OPEN AIR DINING

FIREWORKS DISPLAYS

COURTESY OF UNCLE SAM

OVERNIGHT TENT ACCOMODATIONS

SAFE SANDY BEACHES

MILES & MILES OF HIKING TRAILS

LEECH THERAPY SPAS & SAUNAS

FREE HELICOPTER RIDES

One of the signs that applied to Vietnam
(unknown source who took picture)

One of the Military base camps in Vietnam

CHAPTER 3

Camp Enari

I was flown by C-130 to the airfield around Plei-ku, Vietnam. I was taken to our unit at Camp Enari, which was the 3rd Battalion, 12th Infantry Regiment of the 4th Infantry Division.

I was issued an AR-15 automatic rifle, which I was not familiar with because I had trained on M-14s, so I had about a week to learn the AR-15. I learned how to tear down the AR-15, put it back together, and keeping it oiled so it would function well when it was needed in combat. This rifle was light weight and had twenty round magazines that would fire in seconds.

We were housed in what looked like a big chicken house at Camp Enari. These buildings were about a hundred- fifty feet long and forty feet wide, built on a concrete slab with metal roof and sides with a walk-in door at both ends. We had metal bunk beds with just a mattress, some had mosquito nets and some did not. We slept in our clothes as you never knew if the base would be hit with incoming mortar fire or a ground attack from the enemy.

These buildings held approximately sixty men, at

13

least that is what I recall, could have been more or could have been less. I was only at this location for about a week.

I did not know any of these guys. We were sent there individually and not as a unit, so everyone was a stranger. We all got along great, it came time for us to be sent to our units and I was the only one out of the barracks that was to leave for Charlie Company.

As I got ready to join my unit, Charlie Co. 3/12th Infantry after about a week at Camp Enari, I was issued my ruck (Backpack), two canteens of water, extra pair of socks, poncho (raincoat) that fit over your head and draped to your knees, camo blanket, steel pot (helmet), two hundred rounds of ammunition, and c-rations for one day. From now on, everything I owned would be carried on my back.

I saddled up and walked to the helicopter pad, a half mile from our chicken house. I had never flown in one of these choppers before, so this was another new experience. The chopper was loaded down with cases of ammunition, c-rations, and one orange bag, which I later learned was our life supply line, our mail from home.

There was no room for me to get in the bay of the chopper, so I backed up and sat on the edge of the door with my legs dangling out. This was not a good spot as I was hanging on for dear life. I was the only passenger on this flight. It took a while for the chopper to get the blades turning fast enough for us to lift off, so I was sitting there bouncing with the chopper until there was enough momentum to fly.

We started rising and the higher we got the more scared I was. The thought went through my mind jump-out, jump-out then I thought I would break my legs, so I closed my eyes and hung on for dear life but scared.

We flew right along, and the scenery was a lot different than I was used to at home. The green trees, rice paddies, and red dirt. The ride was about fifteen minutes long and then the chopper started down to land in the middle of nowhere, a dense heavy wooded area, and the next thing I heard was some ping-ping-ping, and the door gunner on my side opened up. That guy could have cut down a tree with that machine gun roaring.

The chopper went up and banked a hard left, my eyes were the size of golf balls, and I was looking to hold on to anything I could, praying for my life that day as I would be praying every day I would be there in Vietnam.

We drew sniper fire so the chopper could not land, and we returned to the airstrip where we had left earlier. The door gunners got out and began inspecting the holes to see how much damage there was, or if it needed major repair. I was still shaking and wondering what I was doing here, I was just eighteen years old at this time.

That night I was assigned guard duty around the airfield; the bunkers were about a hundred feet apart and surrounded the airfield where there were twenty to thirty UH-1 Huey Helicopters setting.

The bunkers consisted of a hole in the ground that was four feet deep and six foot long, with two feet of sandbags on the end bunkers above ground, with a sheet of metal with an additional layer of sandbags on top of that to protect you from incoming mortar rounds. The front and back were open so you could get out either side if needed. The guard shift was usually four hours long and then you would be relieved by another soldier from one of the bunkers.

That night was quite but still you were scared that the enemy was crawling upon you, and you would be killed. It was not only this night, but fourteen more months of praying for your life.

The next day my helicopter, still loaded from the previous day, was getting ready to fly me back to my company. I backed up and sat down on the edge of the chopper as I had the day before, and we took off. I guess my nerves had settled down as this day I was not as fearful as I was the day before.

The sun was shining, the air blowing across my face, the chop, chop, chop sound of the blades taking me to a whole new life where you lived like an animal, but you became best friends with people you did not know.

The chopper landed in this homemade LZ (landing zone). Soldiers without shirts came running toward the chopper unloading the ammunition, C-rations, and the bright orange bag once I was out of the way.

They started loading long green bags which filled the floor of the chopper which I later learned were deceased soldiers and found out I was a replacement for someone who had been killed.

I checked in with the commanding officer, who was dirty, sweaty, and unshaven.

I was assigned to the third platoon, where Sgt John Grasham, from Indiana in a growly voice said, "to do as told and maybe I would live to talk about it." Grasham had been there about three months, so yes, he knew more about what we were doing, and yes, I would gladly do what was asked of me.

This was March 1969; our company had just joined Alpha company which had contacted the enemy and had lost several men the previous day at this location. These soldiers were the ones in the body bags that were loaded on the chopper I arrived in.

Sgt Grasham re-supplied me at the platoon. I received two more canteens of water, a case of c-rations in boxes, which we opened and put in my ruck. As time went on, you would keep the c-rations you liked and the ones you did not like you gave to your buddies that liked them. Ham and Lima beans and canned eggs were my first to get rid of. Each box of c-rations had cigarettes and a chocolate bar, but I did not smoke, so I would give the cigarettes away.

I also strapped on another six bandoleers of ammunition, this gave me close to one-thousand rounds for my rifle, plus ammunition of two hundred rounds for the machine gun, in addition to four hand grenades and a claymore mine that we sat up in front our position at night to set off during a ground attack from the enemy.

The claymore mine was made up of small steel balls packed with C4 explosives which was detonated by a blasting cap which you had control of.

Jim Wytko (WA), Lester Spearman (FL), Bob York (IN), Bob Meyer (MO), Stan Crab (KS), John Wirth (KS), and Don Dixson (KY) made up my squad in third platoon. Meyers, Wirth, Dixson, and I were all new men arriving within a month of each other and every day learning how to live and survive in the mountainous jungle we had just been dropped into.

Being my first day, I was treated like I had been there for six months, the Sergeant sent John Wirth and me outside the perimeter on LP (listening post) duty. The only instruction we received was to shoot anything that moved. The day before, the platoon had walked into an ambush and had lost several men, some of whose bodies were not recovered.

It took a few days to recover from this ambush patrol and account for the lost men, one of the soldiers was listed as MIA, but was later found to be a POW that was released from Hanoi in 1975 (In remembrance of Gail Kerns POW).

I got to know my squad fast, as we all depended on each other to survive. We were never ten feet apart from each other, so we became brothers quickly. We pulled guard together, went on four-man short range patrols together, sometimes eight-to-ten-man ambush patrols. We never stayed in one location. At dark we slept next to each other so we could wake each other if we had enemy movement or if it was time to change guard.

When it became nighttime, all we had was the moon and stars, so we never moved around when set up at night. You would sit there in dead silence, watching and listening for the enemy.

The days passed by, and our company of about one hundred men would patrol through the mountainous jungle that we had been dropped into. Some days we would find old campsites where the enemy had moved out of, and always on the lookout for poisonous snakes, as there were a lot of them in this country. And don't forget the continuous harassment of the mosquitoes.

We were supplied with mosquito repellent but I am not sure that did not draw them. We did take a malaria pill daily and a big orange malaria pill weekly, but they were not a hundred percent effective because several men got malaria.

As we patrolled the jungle, we would take a break at a secure area to cool down from the heat. We were surrounded by trees and thick vegetation so there was no breeze. We would take our backpacks and our steel helmets off to try to cool down, but our clothes were wet from sweat. That is the way we would be until we reached our night location where we could then take off our shirts to dry out.

We would also take off our boots and let our feet dry and air out. Everyone had extra socks, so we would put them on and washed our dirty ones with water from our canteen and lay them out to dry. No one ever got comfortable because a four-man SRP (Short Range Patrol) had to go out at night.

I was one of the lucky ones if you could call it that. I do not recall many nights inside the company perimeter. These four-man SRP were not safe. We were always one click (a click is 0.62 miles) or more from the company, and if we needed help it would be too late by the time they arrived, so you had to be alert and smart about everything you did.

One day while patrolling, we discovered an underground medical facility in a cave on a mountain side. It was a 20x20 foot room with a big flat rock in the middle used as their operating table. It had fresh blood on it, so we were not far away from the enemy soldiers. Over the operating table (rock) there was an opening in the ceiling to let light in, along with candles along the walls. This was a clean room surrounded by rock. We never found any enemy at this location, but you felt like you were standing on the wrong end of the gun, so we moved on without any contact.

Morale was low among the whole company, everybody wanting to go home, as it was extremely hot, and miserable, and most thought this war was a lost cause. No one really knew what we were fighting for, just doing what the United States Army was asking us to do.

As time went on, our clothes continued to be wet with sweat and became well worn, they would re-supply us with clean clothes every four to five weeks, but it was only a matter of time before these rotted off as well, from sweat, or the vines and bush that tore them as we humped through the jungle daily.

Drinking water was flown in weekly and taking a cold shower was a rare happening, if we got a shower every two months it was a special treat, at one point in time we went three months without a shower. During this three-month period, our company set up around a bomb crater filled with water, so I decided I would take a bath, but when I got out, the leeches had covered my body. I got to my bayonet (knife like which aided in hand-to-hand combat) and scraped leeches off my body for what seemed like thirty minutes.

Jim, John, Don, and I went out on a lot of these SRP together; we trusted each other not to fall asleep, and we would give our life to protect each other. These SRP patrols would sometimes be two nights. We would write letters back home or quietly play cards to pass the time. While on patrol we were in radio contact with our company every hour, by sending two squelches to let them know that we were alive and not sleeping.

Camp Enari

Joe taking a bath in 500-pound bomb crater

Entrance to base camp at Camp Enari

Other soldiers taking a bath in crater and after scraping off leeches

CHAPTER 4

Point Man

When our company moved through the jungle, a different platoon would take point (the lead). I had lucked out for about two months, then Sgt Grisham said, "Carter, your turn." This was never the best place to be!

You were reading the compass as it was quite easy to get disoriented in the middle of nowhere. We had to always know our position in case we contacted the enemy and needed artillery support. You better be precise on your location, or you could be the victim. This day I was cutting a path with the machete, reading the compass and map, watching the surroundings, trying to keep moving.

The elements of the jungle, thick brush, the vines grabbing hold of you, watching for the poisonous snakes, it was very humid and hot. I had led the way for about two hours, then I could go no more, I was exhausted from the heat.

Wirth who was right behind me called for the medic (a soldier that we called "Doc"). I sat down and got the pack off my back, the Doc got to me and poured canteen water over my head trying to cool my body temperature down. The company moved on, walking right by Doc and me. As it turned out, there was an enemy bunker complex fifty feet in front of us, but the enemy was gone. We had missed the enemy by approximately three or four days.

John Wirth had taken over as point man. As I got rested and cooled down, it was time for Doc and me to start bringing up the rear of the company.

We were all drowned in sweat, mosquitos buzzing us, clothes were wet, but this was everyday life while in Vietnam. Getting to our location, we set up our perimeter, digging fox holes for the night in hopes of getting some rest. SRP patrols were sent out like every night, and we just hoped that the enemy was not coming up the path behind us that we had just made.

As I mentioned earlier, it was pitch dark here. If the moon shines, that is all the light you had. If you did not get scared at night, there was something wrong with you.

Poisonous snakes were always our enemies, the Bamboo Viper was one of the deadliest. It was about a foot long and green that blended in with the terrain. I was introduced to one the second week of being in country as we were setting up camp for the night. He was killed but was an extremely aggressive snake. There were over twenty-seven different venomous snake varieties, so we were snake hunting as well as fighting the Viet Cong. The python was the largest of the snakes. It could be eighteen to twenty feet long. We also had mean monkeys and tigers to deal with.

July 1969 our company was guarding an artillery company at LZ Penny. My company had about one hundred men, I stayed with my platoon and never really got to know very many soldiers in the company, except for a few. This day was no different than the others, I was going out on a four-man SRP patrol that night.

I always felt safer on these than staying in the perimeter but now being older and looking back, I realized I was an easy target for the enemy on these patrols. I lived to talk about it, so guess the good LORD was watching over me. At midnight things seemed quiet until one of our SRP patrols started burning up the radio.

I woke my patrol up and we were all looking for the unknown. The other SRP patrol had a two-week new man, which woke up scared and ran off in the dark. His patrol later found him dead, a tiger had killed him and dragged him off. The men scared the tiger away and were able to recover the body. There was nothing that anyone could do, but it did make you sleep with an eye opened, as I had done since day one of arriving in Vietnam.

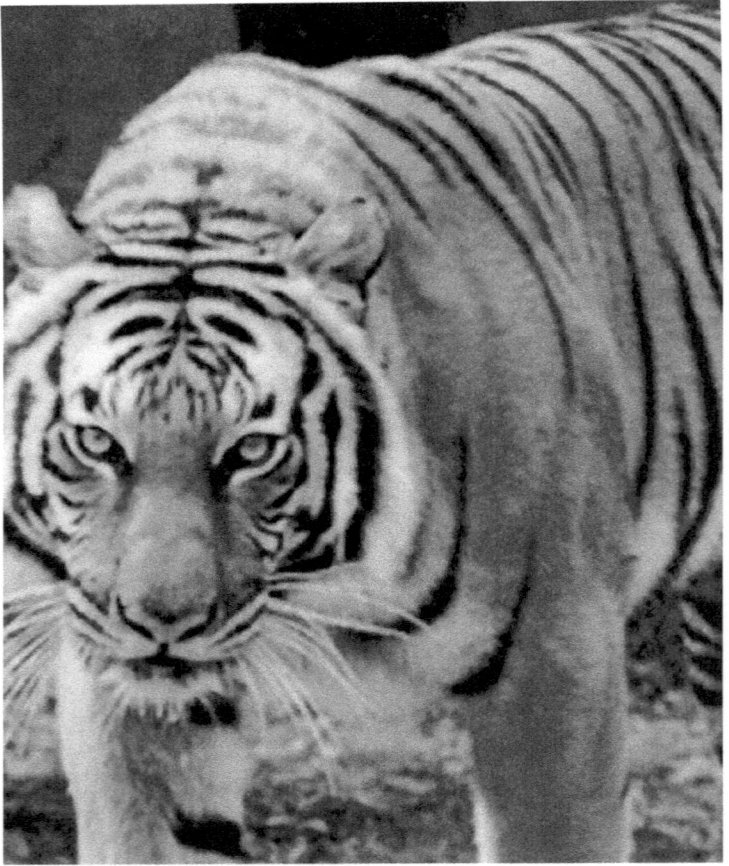

One of the enemies of the Jungle (Picture taken by unknown source)

One of the many snakes we encountered in Vietnam

SRP Patrol Dixson, Wytko, Me , Wirth

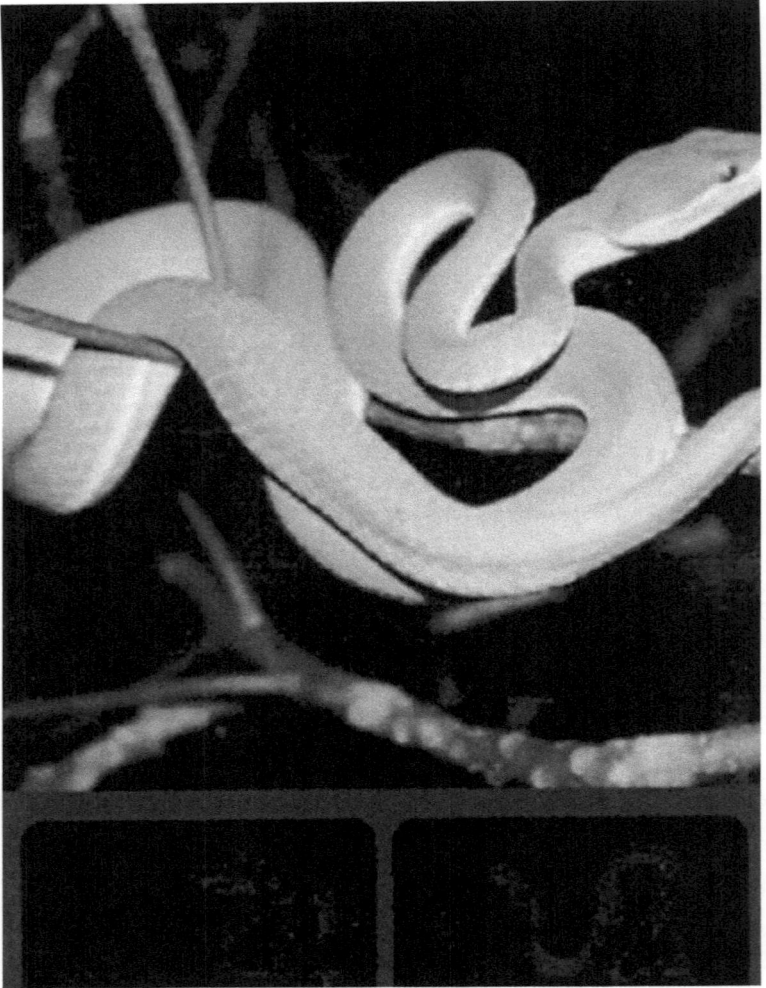

Bamboo Viper (2 stepper)

CHAPTER 5

LZ Cherokee

June 17,1969 was no different than any other day, just counting the days you had left in this miserable country of Vietnam. Our company had been moved to a flat ground location, northwest of Pleiku to destroy enemy forces and prevent enemy rockets and ground attacks on Pleiku.

A bulldozer had been flown in by helicopter two days earlier to clear brush and trees from our perimeter two hundred feet around so we could see the enemy if we were going to be attacked. Engineers had come in and put three rolls of razor-sharp concertina wire around our perimeter.

Our company never stayed in one location for more than three days at a time, but at this location we were there approximately six days. Every day we sent out SRP patrols, sometimes for two days at a time. I was either in charge of taking one out or was a member of the patrol.

Our company was short of men all the time, so by each platoon sending out a SRP (that would be 16 men from the company), that left about forty-five to fifty men inside the perimeter. A company was supposed to be one-hundred and twenty men strong, but we struggled to have a hundred men in our company the whole time I was there.

June 17, 1969, my SRP with John Wirth, Bob York, and Don Dixson left the perimeter about 4 PM. Nothing seemed unusual as we arrived at our assigned location about thirty minutes away. We did not find a good place to set up, the land was open, no towering trees, just small clumps of vegetation.

Being a one-night operation, we elected to set off a trail where we found small brush cover where we could see in front of us and behind us. We radioed our status to the company, and the four of us laid down next to each other so when it was time to stand watch, we did not have to move around, or if someone snored or had a bad dream we could punch them to wake them up to quiet them down.

Our Company had a ground attack that night at 12:20 am, my SRP was awake, and we could hear rocket explosions, and gunfire. Then at 1:12 am, twenty enemy soldiers walked by us. We were not in an ideal spot as we had no real cover to hide behind, so we had to let these enemy soldiers pass by, which was a good idea because five minutes later approximately seventy -five enemy soldiers walked past us. Had we killed the first twenty enemy soldiers, we would have had seventy-five enemy soldiers surrounding this four-man SRP.

At 1:28 am, Spooky (AC-47 aircraft) was called in and flew around the company perimeter, with mini guns laying down fire all around the perimeter for thirty minutes. Spooky was then released from the area and returned to Pleiku to re-arm and stand-by for further support. We had marked our position with strobe lights so that the aircraft would know our position. Being scared in the pitch dark of night, we were all praying for our lives and our brothers who were fighting in the attack.

The next day, June 18, at daylight the second platoon swept the area and recovered thirty-two grenades, twenty-three satchel charges, and hundreds of spent 7.62 shell casings, and mortar rounds. One North Vietnamese body was found, and Jim Wytko was given the order to bury him. The enemy would always drag off the KIA's (killed in action), and wounded, so it was hard to get an accurate body count after a battle.

Our SRP returned to the company where I discovered my side of the perimeter had taken a hard blow from the attack, I lost eleven brothers that night, eight WIA (wounded in action), and three killed. This battle has never been recorded in any of the history books about the Vietnam war, because it was deemed an insignificant battle, but will live vividly forever in my mind, and those who fought that night.

In Memory of the following:
SGT James A Cowan Jr.
SGT Barry R Grulke
SP4 Dieter E Willert

Aftermath of a ground attack.

Joe and John Wirth

Me on SRP Patrol

CHAPTER 6

Dear John Letter

I had been in Vietnam about six months. In July 1969, our company was guarding Firebase LZ Penny, being short on men from the ground attack from the month of June.

I had entered the Army at 18 years of age, so I had left a high school sweetheart at home. I carried her picture and she wrote letters weekly, keeping me up on all the small-town news. Her letters always smelled of perfume, so I would have to pass the envelope around my squad so they could get a good whiff of it before I got to open it up.

Getting mail from home was like receiving a Christmas gift on Christmas day. It always lifted your spirits and made you feel good inside.

My platoon was sending out SRP patrols every day, always on the lookout for suspicious activity. Mail call came around before the SRP was getting ready to leave the perimeter. The Sergeant was passing out mail. When he handed me mine; I was happy to see that letter from my girlfriend, but this time it did not smell of her perfume. I sat on the bunker, opened the letter, and began reading it over, and over, and over again.

I told my buddies about my Dear John letter; she had found a new boyfriend and was going to quit writing to me. I started reminiscing about back home, and the good times we had. I was 19 years old at the time, my life was at the point of do or die. You had to focus on everyday life, I was fighting to survive every day in the jungle and, sure this bothered me, but I had a job to do over here.

I went out on the SRP patrol that night as I knew I would not sleep well, thinking about that letter and the plans we had for when I returned home. About 2 am, another SRP patrol had some movement, so you could never let your guard down. This was an instance of where you had to be on the lookout for the enemy but also the enemies of the jungle such as tigers, which on this night had attacked a soldier on another SRP patrol.

CHAPTER 7

Monsoons

We were still in the mountainous jungle, moving from one location to another, never staying in one place. We would make brief contact with the enemy and they would disappear. If jungle fighting in this mountainous terrain was not enough, the monsoon season began to set in.

It rained day and night, night and day for twenty-eight days; we were drowned rats daily. We had our raincoats on, but our clothes were wet, and never had a chance to dry out. We had to sleep with our backs to a tree as the water puddled everywhere. Our feet were soaked, but we kept on patrolling the jungle. We tried to keep our rifles dry by keeping them under our poncho (raincoat). We also tied baggies, or anything, over our gun barrels to keep them dry. This was our only way to keep the rain from running down our gun barrels, causing our rifles not to function when we needed them.

I counted the days; I was soaked twenty-three days out of the twenty-eight days. Those few days we would see the sun, but it continued to rain. We continued to patrol the jungle, just like any other day, as SRPs went out, ambush patrols went out, the rain did not stop anything… but the helicopters from flying, which meant no mail, no c-rations, no ammunition or fresh water, so we had to be mindful of what we did have.

Of the fourteen months I was there, we always had enough c-rations, but water would get low at times. We never ran out, but it was limited. We could not fill our canteens out of the rivers, for fear of dead people or animals up stream, plus this was how the Vietnamese people would bathe and wash their clothes. Hepatis from drinking this water could be contracted, making you extremely sick.

The monsoons lasted about a month. It was day twenty-eight, I became sick. It was extremely hard for a soldier to get out of the field to go see a doctor at base camp, as the company was always short on men. Our platoon medic was aware of my condition, and went with me to see the captain, and assured him I needed medical attention. The captain did not want me to leave, but we had a helicopter coming in, so he gave me the go ahead to leave.

I flew into Pleiku and checked into a Mash Unit near the hospital. After getting checked over, they put me on an Army cot at the back of the tent, probably because I stunk from sweat and dirty clothes. I never got to take a shower, or clean clothes or even offered any. They gave me medication and thought I would be out of there the next day, but boy did I surprise them.

About midnight I woke up vomiting all over their concrete floor, which did not go over well. The next day they transported me down to the airfield and put me on a C130. I had no idea where I was headed, and by this time, I was extremely sick and did not care. We landed at Quin Nhon, where I was taken to a hospital. It was here they took my dirty, sweaty, stinky clothes, and I got my first hot shower in almost seven months (in the country), where I was cleaned up.

After taking my blood, they escorted me down the hospital hallway to my bed. I was starting to feel better already, as this was my first bed with white sheets since I left home in February 1969. Everything else had been cots, iron beds with three-inch mattresses, and mostly the ground in the jungle.

Getting to my bed, John Wirth from my squad, and our platoon medic Doc Stofferson were right next to me. They were both in the field when I left, so I was incredibly surprised to see them. They beat me to the hospital; they both had malaria.

They were there four or five days and left, later learning they were sent to Cam Rahn Bay for thirty days for rest and sunshine. I was not as lucky, as I was diagnosed with Typhus, caused by being wet for a lengthy period. So, I was confined to the hospital for ten days.

During my hospital stay, I had written to my mom and dad telling them only I was in the hospital, but the doctor said I would be okay and said I would be there for a while, which was about ten days. Being in the Infantry, I should have explained more in my letter to them about why I was in the hospital instead of worrying them and having them imagine the worst. I later wrote another letter explaining more about my hospital stay.

I was released from the hospital around October 4, 1969 and transported by truck back to Pleiku where I was choppered out to my company. Returning to my company, I was sent out on ambush patrols and four-man SRP, nothing any different than before my hospital stay, except two men from our platoon had gone home.

Jim Wytko had gotten sick in July and was sent to base camp, and I never saw him again until 2009 when I went to see him at his home in Bukley, Washington. Jim went home in January 1970.

Bob York did his tour of twelve months and went home in September 1969. Lester Spearman did his tour and went home in October 1969. SGT Grasham's tour ended in January 1970.

C-rations, normally carried 1 week supply on your back

CHAPTER 8

I'm Hit!

Halloween night word came down that we had a cease fire. To the infantry this does not mean a thing. It is either your life or the enemy's life, so we were always on guard or knew what was going on in our surroundings.

I was inside the perimeter; the enemy was probing us to see where our weakest point was. The company was on full alert, everybody was awake and on the line, waiting and watching, listening for any sound that would give the enemies position away.

Our Forward Observer was getting our position relayed to an Artillery support base, so we could have support when needed.

Our bunker was small, three people could get in it, so I laid down behind it, as the first artillery spotter round was fired in our direction. We were all looking and we could hear the whistling sound coming in our direction and expecting it to hit in front of us, but to our surprise it hit high in the trees behind our bunker. As soon as the round exploded, I immediately shouted for the medic, "I'm hit, I'm hit!"

When he got to me, we wrestled my shirt off, but could not find any blood, so then I dropped my drawers, but still no blood. This round had hit a tree and riddled the bark in pieces, and this is what had covered my body. Between the concussion, and the fragments from the tree that had covered my body, and the bouncing of my body on the ground, it was just another scary night in hell. The night calmed down after this, and we did not have any more probing, or movement, but everyone was awake and on guard all night.

The days went on, getting clean clothes once a month, mail, and c-rations every two to three days, we kept moving, patrolling the jungle. Thanksgiving was coming up, but for us it was just another day in the bush. Base camp mess hall sent out a hot meal of meat, instant potatoes with gravy, vegetables, and quarts of milk. The milk had sat out all day in the heat and was clabbered, but we did not get it often, so we drank it anyway. The meat, I was not sure of, but I was thinking it could have been lamb. Our taste buds were stuck on c-rations, so any hot meal was a buffet for us.

Our company continued to patrol the mountainous jungle region searching for the enemy. Some days were quiet, and some days we would have contact with the enemy and they would suddenly disappear.

One day our patrol caught the enemy taking a

bath in a small stream; they had not posted a guard. They were able to get to their rifles and put up small fights, but our patrol outnumbered them. The fire fight lasted five minutes or so but had your adrenaline going, and you were scared just like every time you made contact with the enemy.

Me setting up camp for the evening

Me at one of our many locations

Blowing up trees with explosives to let a helicopter in

Bob York, John Grasham, Bob Myers

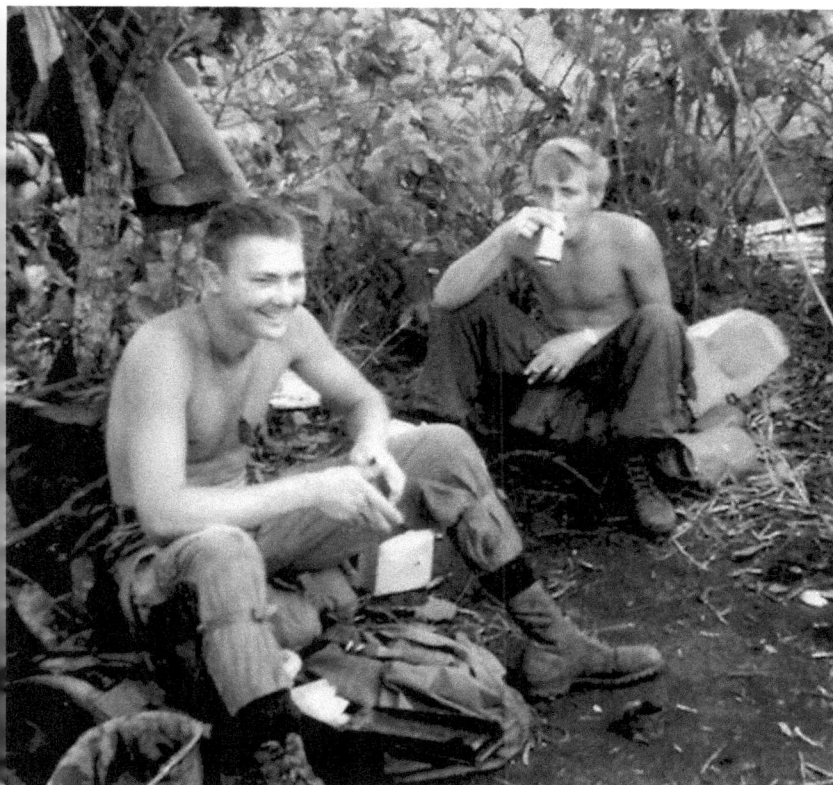

Finally getting some down time in the field

CHAPTER 9

Christmas

The days were passing slowly and Christmas of 1969 was coming up. We were given a Red Cross package containing shaving cream, razor, toothbrush, and toothpaste, writing paper and a pen, and received a package from home with homemade cookies. We got another hot meal out of base camp, but I know it was not turkey. Cease fire had come down to us for the holiday, but like every day, your rifle took every step you took. I was inside the perimeter, sleeping before my guard duty. It was midnight, and I was shaken by one of my squad members to wake me up. Gooks inside the perimeter!

This was not good, how in the hell did they get in? Did somebody fall asleep? We were guarding in front of the line, we were watching our back, a lot of movement going on inside the perimeter. I was hugging the ground hoping no one would get hit by enemy fire or friendly fire. It was finally discovered that wild monkeys were swinging in the trees and dropping down inside our perimeter and then swinging their way back out. Do not know how many times my heart felt like it stopped this night, but another scary night in the jungle.

January 1970, I had been in Vietnam for nine months, so I was getting to be one of the older people, so I got promoted to Spec 4 and later to Sergeant, with time and experience. All my squad had DEROS'd (Date Estimated Return from OverSeas) out of country, doing their twelve-month tour and getting sent back to the United States.

John Wirth was equal in time with me, but somewhere we got split up. He was running patrols out of Camp Enari, and I was assigned to the bunker guard around Camp Enari. We still saw each other and when he was short on men to go out on patrol, I would volunteer to go out with him. We searched villages and tried to keep the local villagers happy by giving away c-rations and chocolate bars when we had them.

It was here that the Chief of a village insisted I try his homemade rice wine, and a couple of days later I had Hepatitis and dysentery. Not a fair trade for trying to be friendly! So back to the hospital I went for lots of tests, medication, and a couple of weeks of light duty.

March came around, time for me to go home. My twelve months have come and gone. I was a nineteen-year-old Sergeant pulling perimeter guard at Camp Enari. This was a decent job and a fairly safe job. The Army had not treated us well in Vietnam, and I had six months to do stateside when I returned, so I decided to gamble and extend my Vietnam tour forty more days to get an early out from the Army.

I did odd jobs around Camp Enari until it was turned over to the South Vietnamese Army and we moved on to Camp Radcliff, near Kontum. The last part of April I was sent back to my company, which were all new men to me. Being a Sergeant, I did not go out on patrols again, but the Vietnam conflict took a different turn... we were being moved into Cambodia.

American troops had never operated in this country, according to Military records, but I knew soldiers that had already been inserted or ran patrols in that country. Cambodia was the supply route for the North Vietnamese to move men and supplies from North to South Vietnam.

Word had come down that our company was moving into Cambodia. This was the last of April 1970, I had five days left on my tour before I was to go home, and here I sat on the chopper pad waiting to be air lifted and inserted into the hostile area that we knew there would be a huge resistance.

Everyone was scared, even though they did not admit it. I had asked to be sent back to base camp to start my DEROS out of country, but that was denied.

They needed all the manpower available as this was uncharted territory.

We were resupplied with new water, c-rations, and ammo to last us a week. Early in the morning, the helicopters started coming to pick us up. I was in the third group going out. There were approximately ten helicopters that hauled sixty men at one move. The first airlift loaded and took off. We waited on the helicopter pad for them to return and pick us up. The first airlift started to return, still loaded. The first helicopter in Cambodia nearing our drop was shot down, with an RPG (rocket propelled grenade). Everyone was killed so the remaining helicopters returned, and the operation was shut down.

The next day I was sent to base camp to turn in my rifle and gear and sent to Cam Rahn Bay to be processed back to Fort Lewis, Washington. My fourteen-month tour in Vietnam was over with lots of good memories and lots of bad, unforgettable memories.

When my time had come to leave, I wanted to go home, but I thought I was letting my buddies down by leaving, so I had mixed emotions.

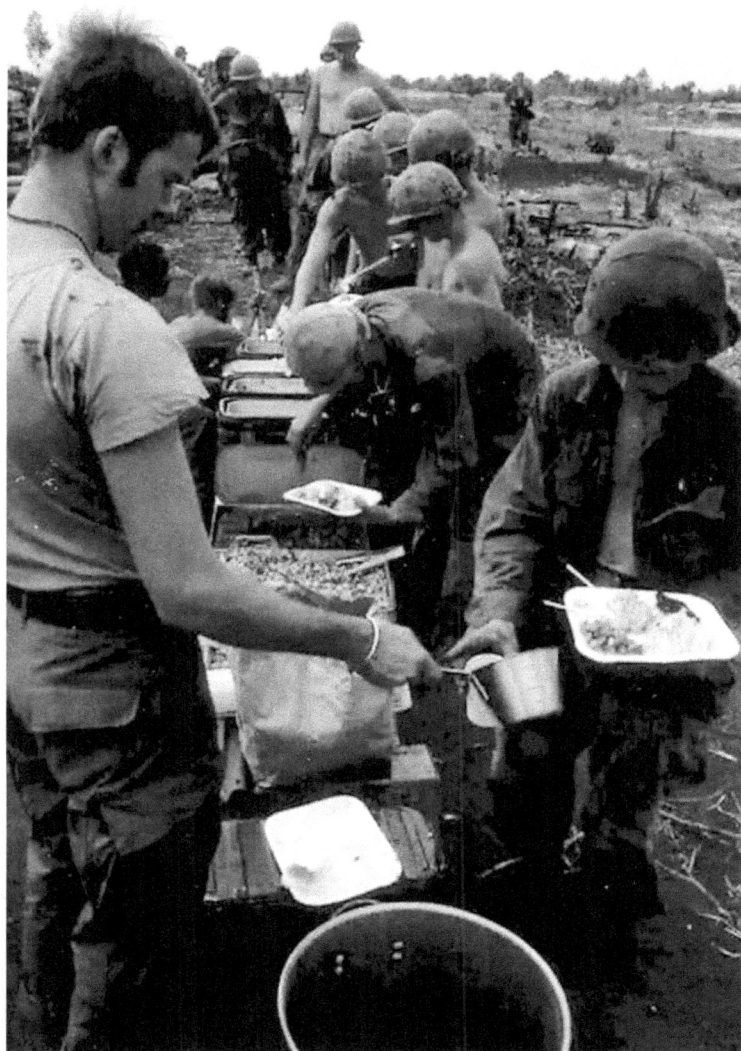

Christmas Meal (Unknown source took picture)

Chopper coming in to move us to new location (unknown source who took picture)

Arriving at new location (unknown who took picture)

*One way we were moved to new location (picture taken by un-
known source)*

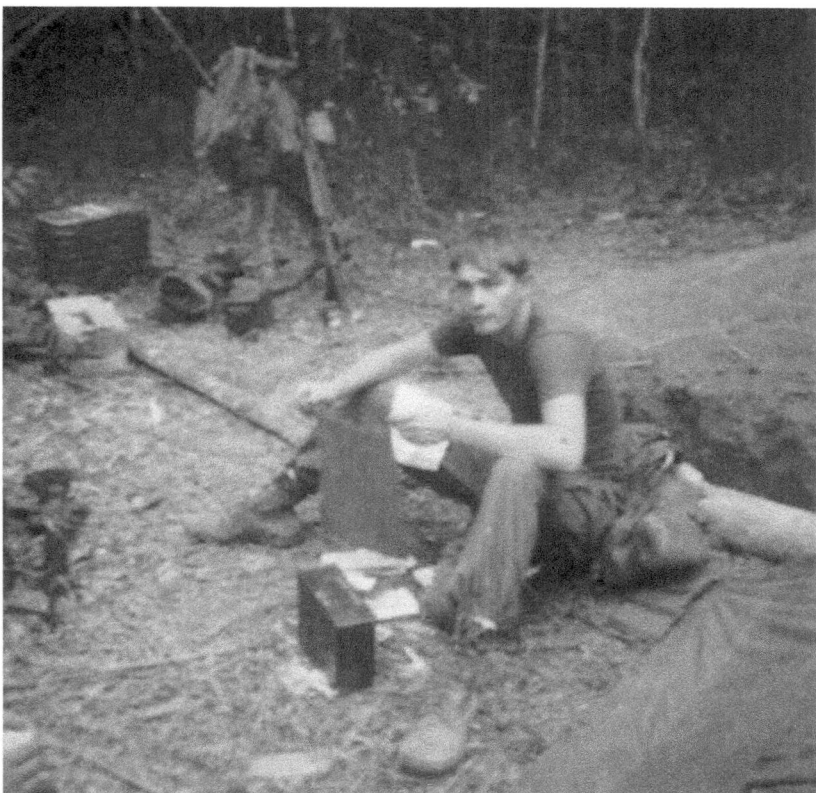

Me with my Red Cross package Christmas Day

One of the villages we searched

www.ingramcontent.com/pod-product-compliance
Lightning Source LLC
Chambersburg PA
CBHW041209220326
41597CB00030BA/5251